PATTERNABLES

A PATTERN BLOCK ACTIVITY BOOK

Developed by   Susan Sehi-Smith

Drawings by   Colette Cappetto
Amy Kemmerer
& Stephen Smith

©Copyright, 1990. Learning Resources
Deerfield, Illinois 60015

INTRODUCTION

Notes to the Teacher

Manipulative materials are a very important concrete tool in helping students improve their understanding of mathematical concepts they are learning. Not only can manipulatives be used effectively to introduce new concepts but they are important in reinforcing and remediating acquired skills.

This book covers a wide range of activities that can be used by students of many different ability levels. It is important to allow students time to explore patterning and to study the relationships of shape and color in the formation of designs. Although the tasks range from very simple to somewhat complex, limitation should not be placed upon children based on grade level.

Students will enjoy making their own patterns as well as solving the puzzles presented in this book. The teacher should extend these activities to include discussions about counting, number operations, fractions, descriptive language and art. Children should be allowed to enjoy this manipulative in a creative, non-competitive environment. Given the freedom to explore ideas, children will become familiar with formal concepts in an intuitive way, long before they are able to formulate and apply these discoveries.

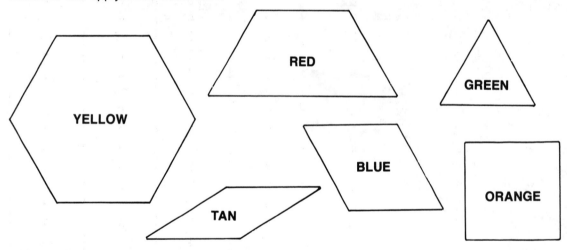

Each set of Pattern Blocks consists of six plastic shapes, each in a unique color. 250 blocks come in a reusable storage bucket in the following assortment: 50 green triangles, 50 blue parallelograms (rhombi), 50 red trapezoids, 25 orange squares, 25 yellow hexagons, and 50 tan diamonds (also a rhombus). Each side is one inch long, except for the trapezoid which has a two inch base. For individual use and for small groups an introductory set of 124 Pattern Blocks is available.

The Pattern Blocks and this book are designed to be used in a variety of classroom settings: in a laboratory situation where the tasks are introduced by the teacher before the children are allowed to work independently; in a regular math class activity with small groups of students who can work together; in learning stations where groups of students rotate from one exploratory activity to another; and in a structured, formal situation where the teacher explains the task and the entire class works through it, together. Teachers should be sure to select activities that are appropriate for their students and to allow sufficient work time.

How to Use this Chart

When first introducing Patternable exercises to very young students, it is a good idea to present the exercise and the exact assortment of Pattern Blocks necessary to complete the picture. This approach reinforces the concepts of shape matching and one-to-one correspondence. As they develop a familiarity with the materials, encourage children to read the chart and sort out the proper Pattern Block combination to complete a given exercise.

Construct a blank copy of the chart or use strips of blank paper to cover the numbers in each column of the chart below. Ask advanced students to work an assigned exercise and fill-in the columns with their results. Initiate a discussion about the findings. Discovery topics should include: Are there more than one set of "correct" numbers? Why might the numbers vary from the printed chart? What is the most number of pieces that can be used to complete the exercise? What is the least number of pieces needed?

Have students write down their estimates of the numbers and types of Pattern Blocks necessary to build an exercise *before* they solve the puzzle. Record the actual solutions after construction and compare to the predicted numbers. Discuss the findings.

EXERCISE NAME	⬡	⬢	◆	▲	■	◢	TOTAL PIECES
Flower Pot	1	2	—	9	6	5	23
Penguin	2	2	6	3	1	1	15
Gift Box	—	—	10	—	4	9	23
Candle	3	3	2	5	1	4	18
Daisy	6	—	3	8	5	—	22
Elephant	3	2	2	—	4	2	13
Dog	2	1	8	1	5	—	17
Turkey	2	6	4	7	4	1	24
Ice Cream	5	7	5	1	0	4	22
Kite	3	5	6	5	4	—	23
Tooth Care	1	2	7	9	3	2	24
Car	3	6	6	4	—	—	19
Spider	—	2	6	2	14	—	24
Telephone	3	10	2	5	—	5	25
Giraffe	3	8	9	7	4	2	33
Crab	1	—	4	—	18	—	23
Snowman	1	10	6	6	6	4	33
Camel	3	7	5	2	1	1	19
Mother Bird	3	4	9	—	7	—	23
Dinosaur	6	4	1	1	2	1	24
Boat	5	6	6	5	4	7	33
Lamp	6	6	10	10	1	2	35
Rocket	2	12	5	2	3	2	26
Key	5	4	17	4	—	—	30
Tree	—	9	6	9	—	1	25
Kangaroo	2	2	4	2	7	3	20
Cat	3	7	4	4	3	—	21
Lamb	5	1	6	5	1	6	24
Train	3	6	7	7	2	5	30
Guitar	1	10	1	6	6	4	28
Bunny	4	4	2	10	5	3	28
Fish	2	8	3	2	11	—	26
Butterfly	4	3	8	7	10	9	41
Turtle	3	7	17	4	1	—	32
Snail	1	4	4	8	11	9	37
Frog	1	3	—	4	6	3	17
Tools	1	9	8	3	6	5	32
House	—	4	8	—	8	8	28
Clothes	2	8	5	10	9	4	38

Cover each of these shapes with a matching Pattern Block.

Cover each of these shapes with Pattern Block pieces.

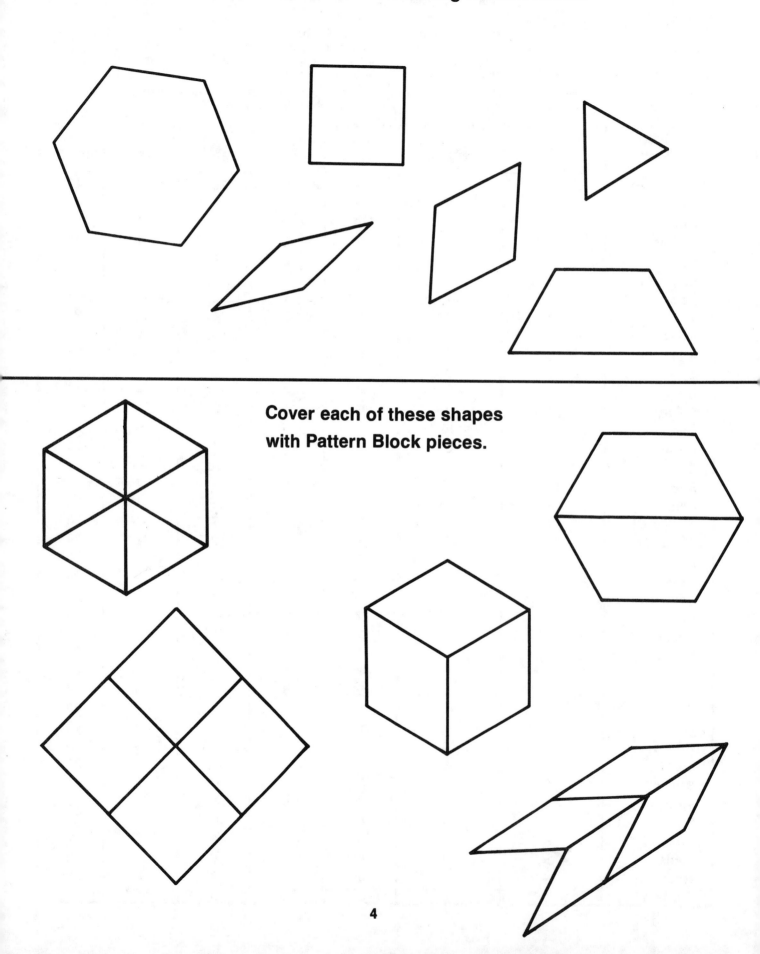

4

Use Pattern Blocks to make these Flowers.

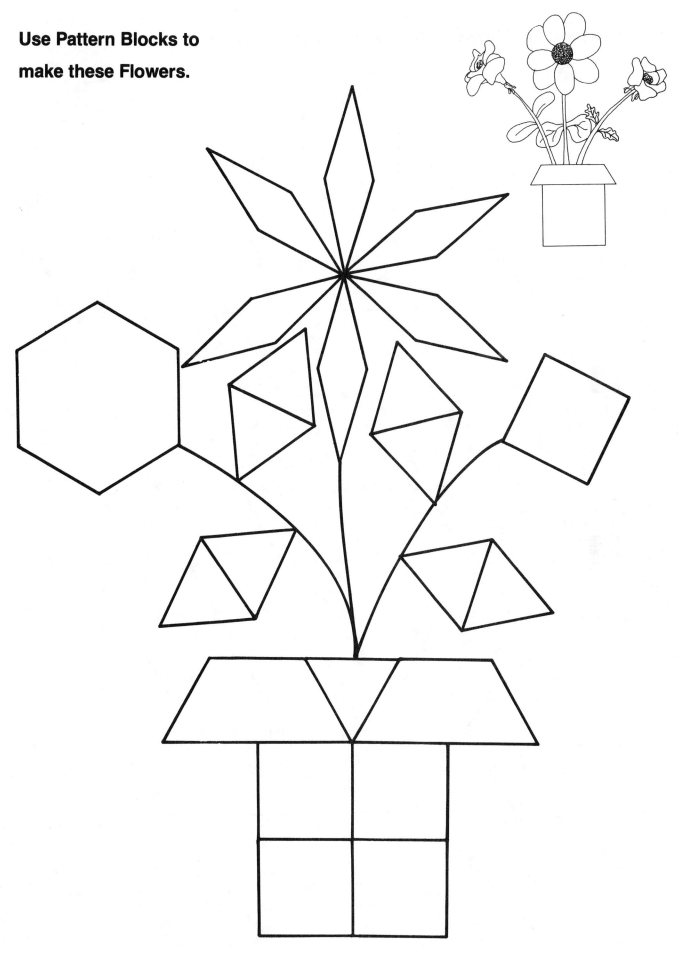

5

Use Pattern Blocks to make this Penguin.

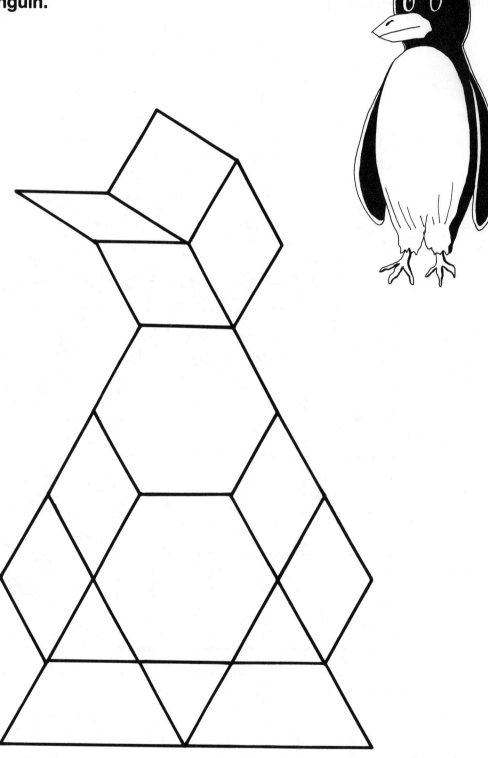

**Make this Present
using Pattern Blocks.**

Can you make this Candle using Pattern Blocks?

**You need Pattern Blocks
to make this Flower.**

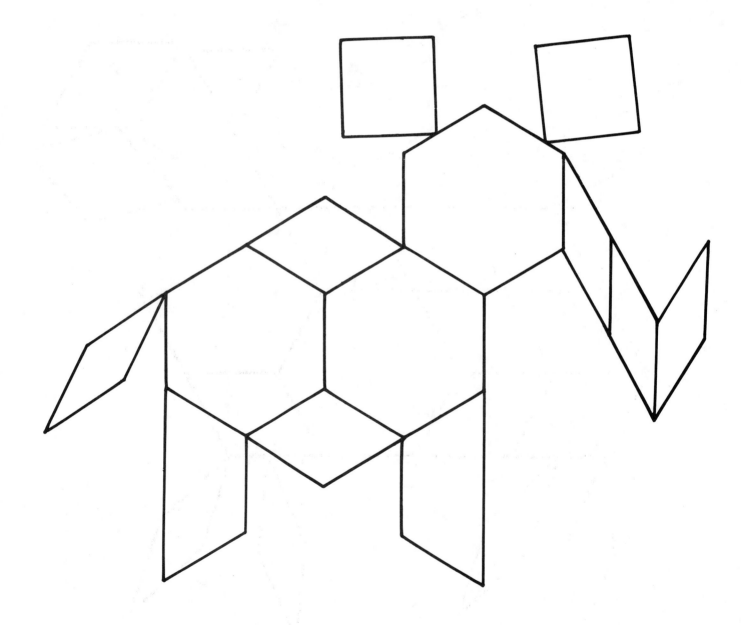

10

**Make a Dog using
Pattern Blocks.**

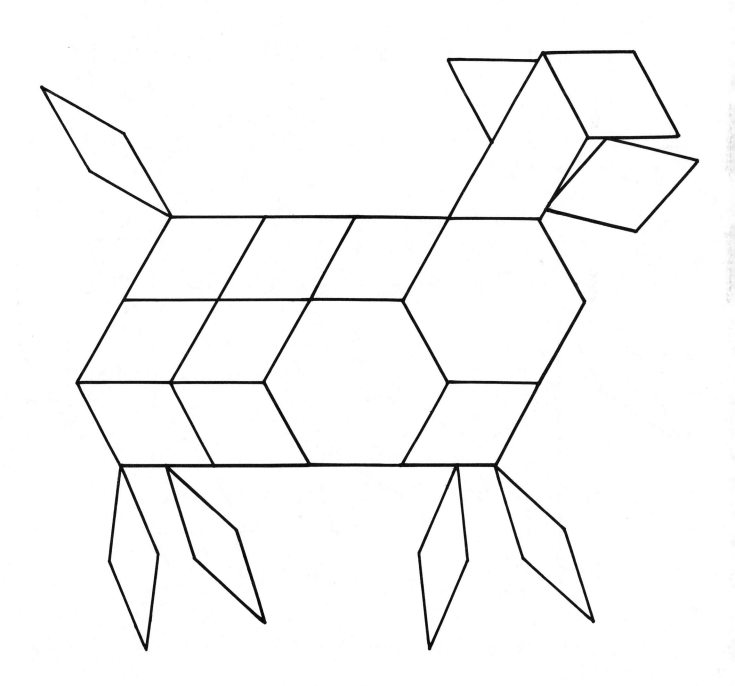

Cover these shapes with Pattern Blocks.

Trace the shapes on paper.

Draw a face on each so that it looks like an animal head.

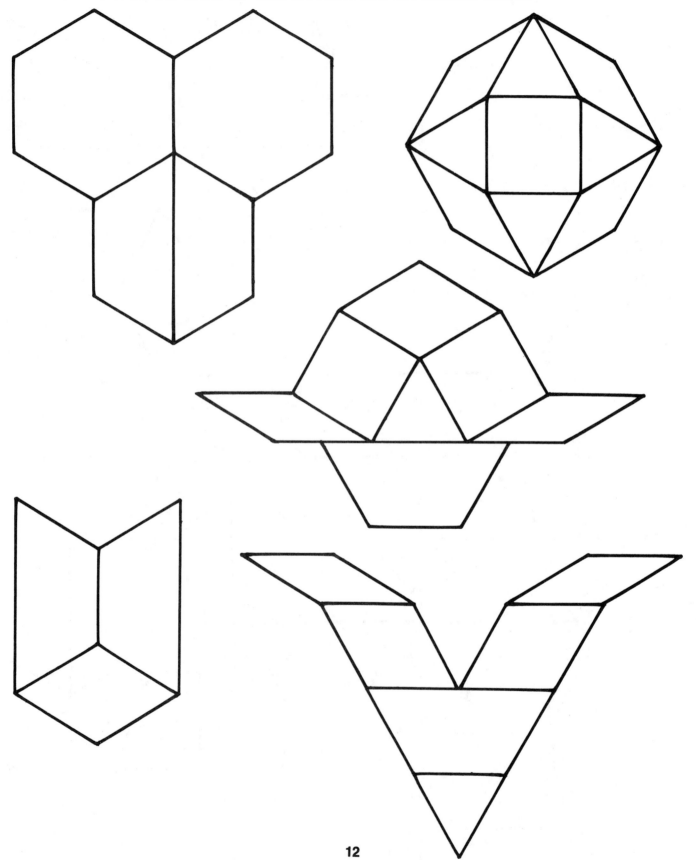

More Fun With Animal Heads.

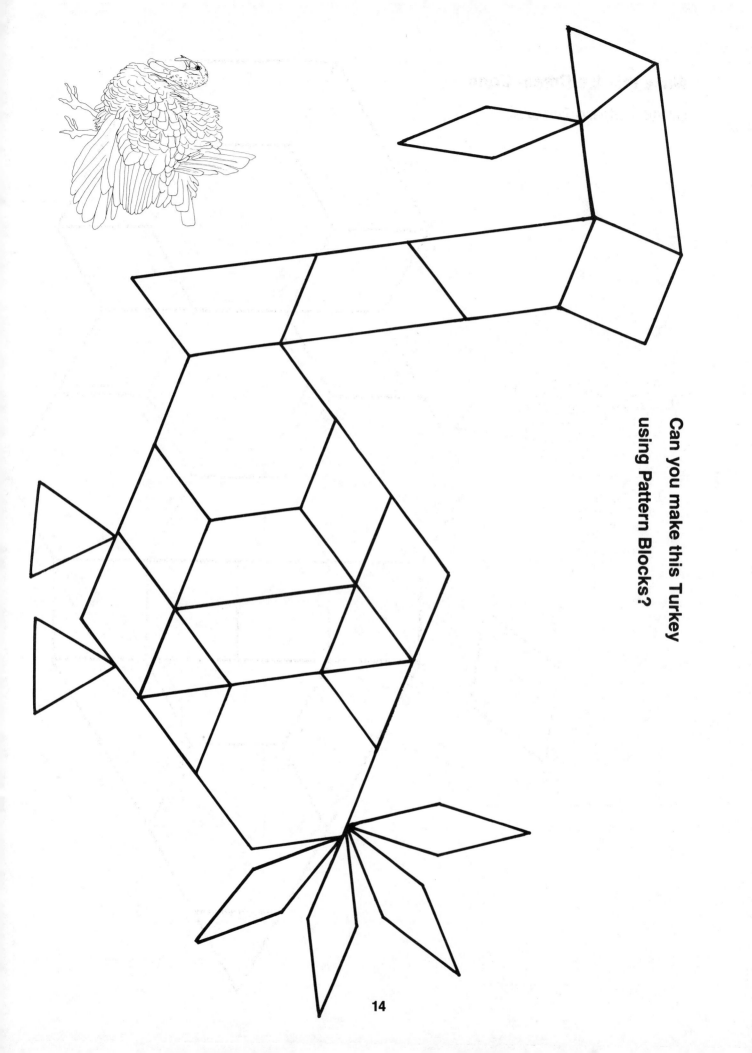

Can you make this Turkey
using Pattern Blocks?

Make this Ice Cream Cone using Pattern Blocks.

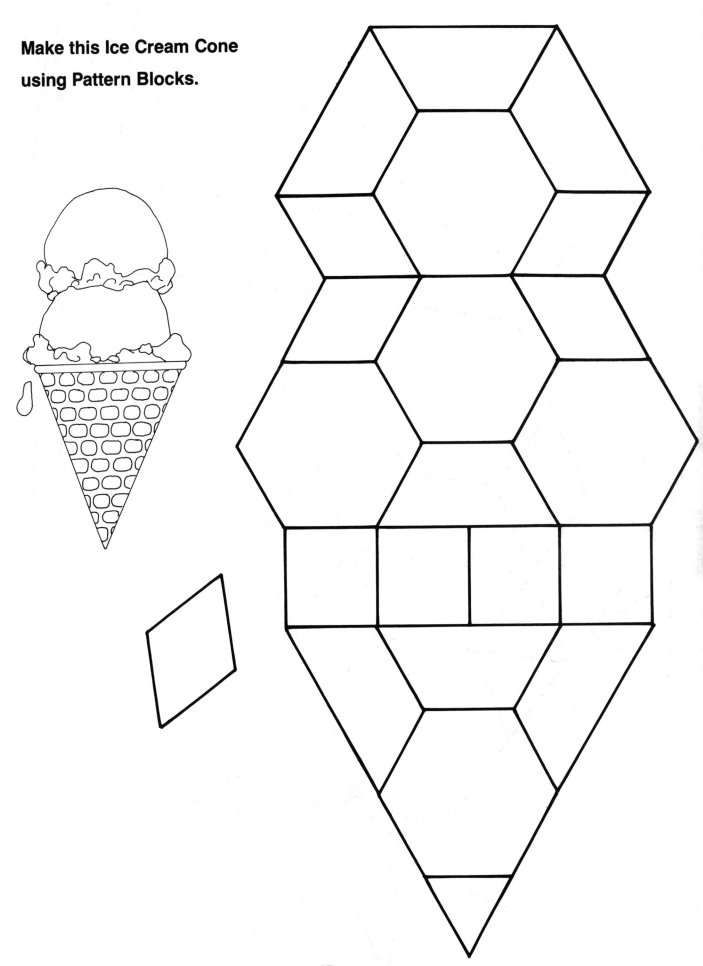

**Use Pattern Blocks
to make this Kite.**

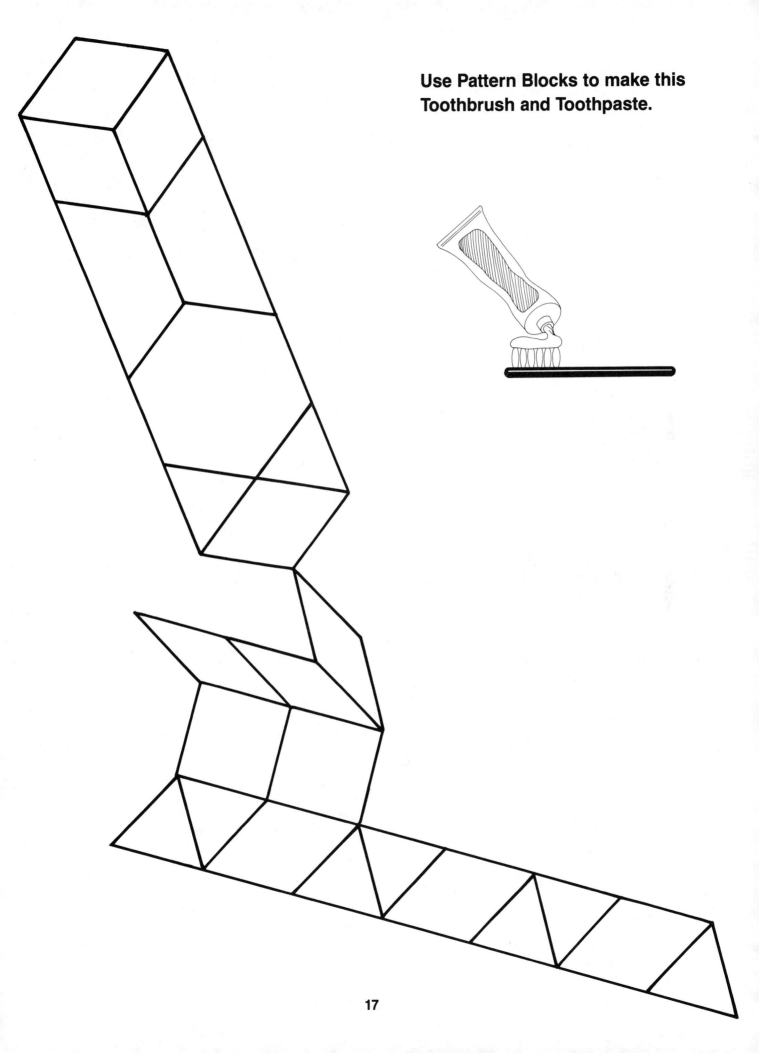

Use Pattern Blocks to make this Toothbrush and Toothpaste.

Make a Car using Pattern Blocks

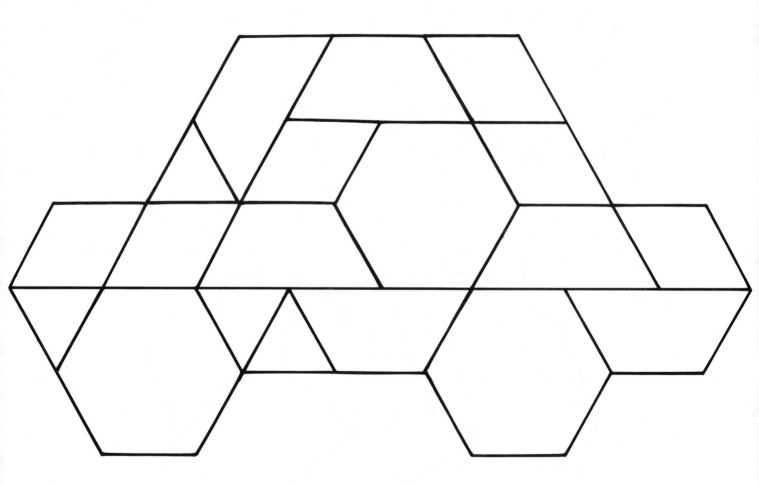

Make a Spider using Pattern Blocks.

19

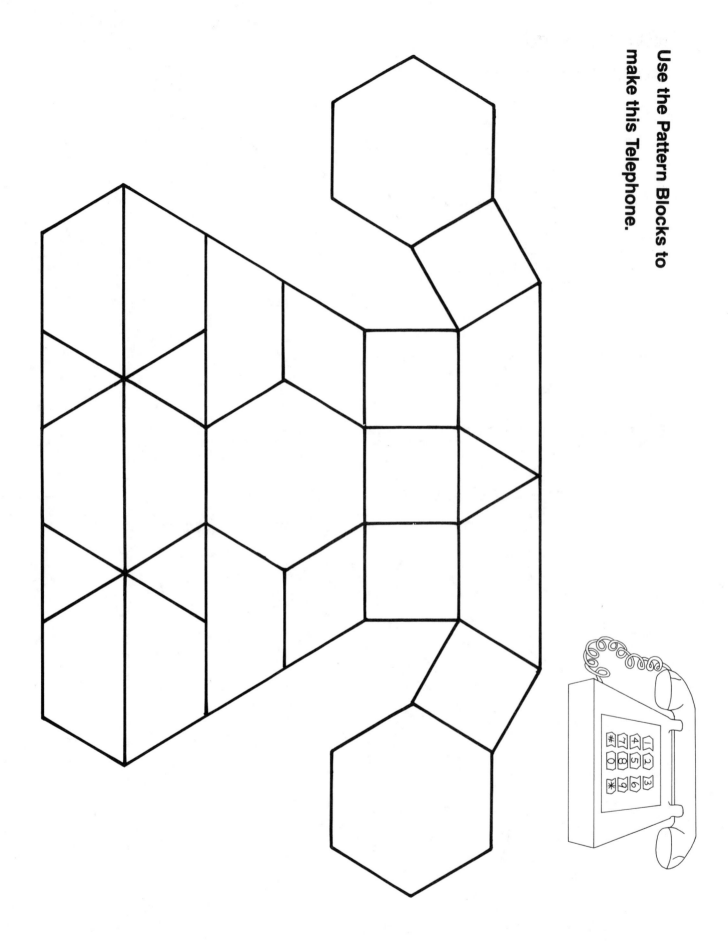

20

**Make this Giraffe
using Pattern Blocks.**

21

Cover these shapes with Pattern Blocks.

Trace the shapes on paper.

Try making other designs that look like people.

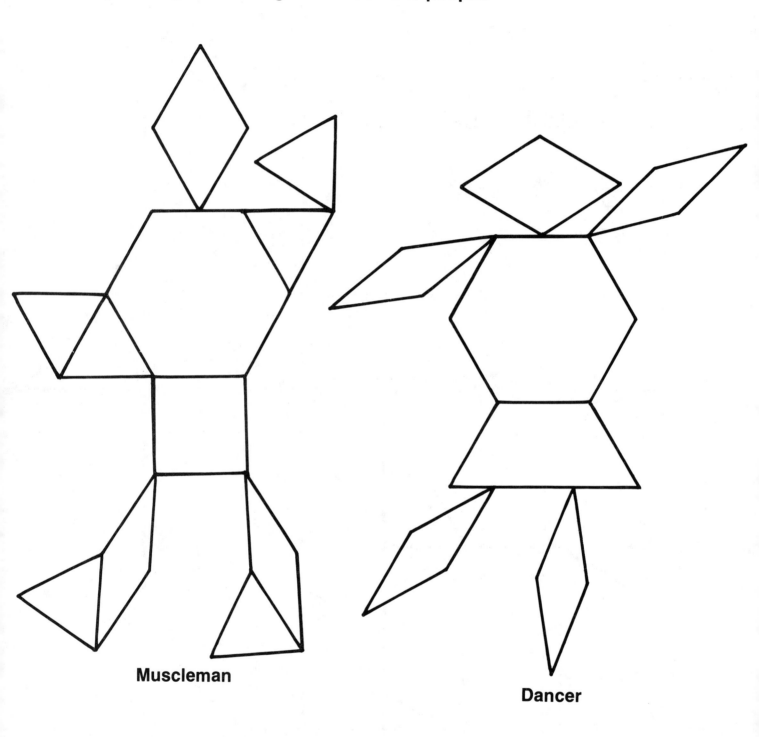

Muscleman

Dancer

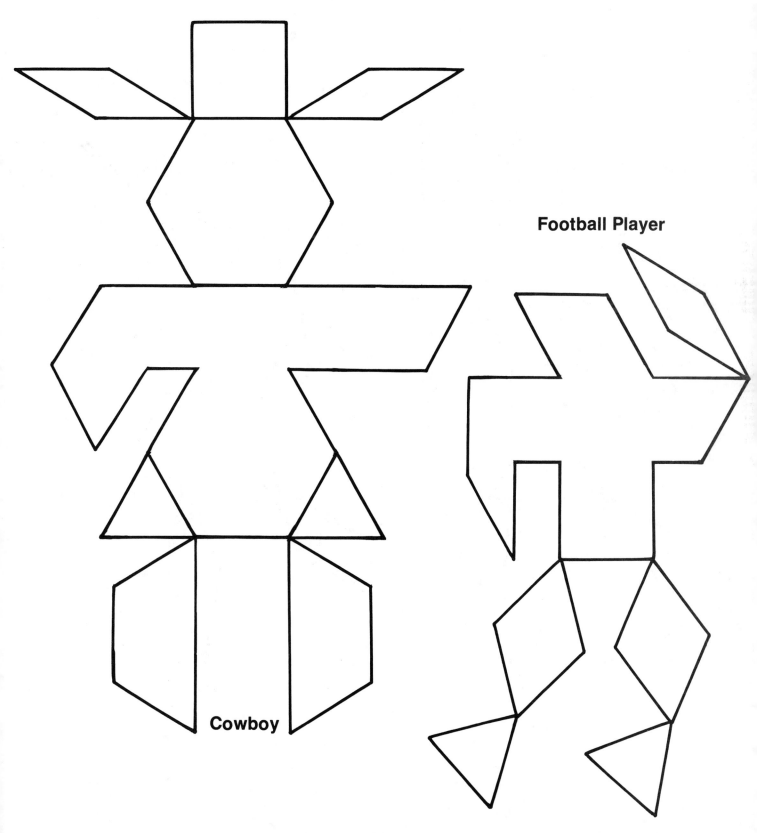

Football Player

Cowboy

Make a Crab
using Pattern Blocks.

Clue: *Use 18 tan rhombi.*

Make this Snowman using Pattern Blocks.

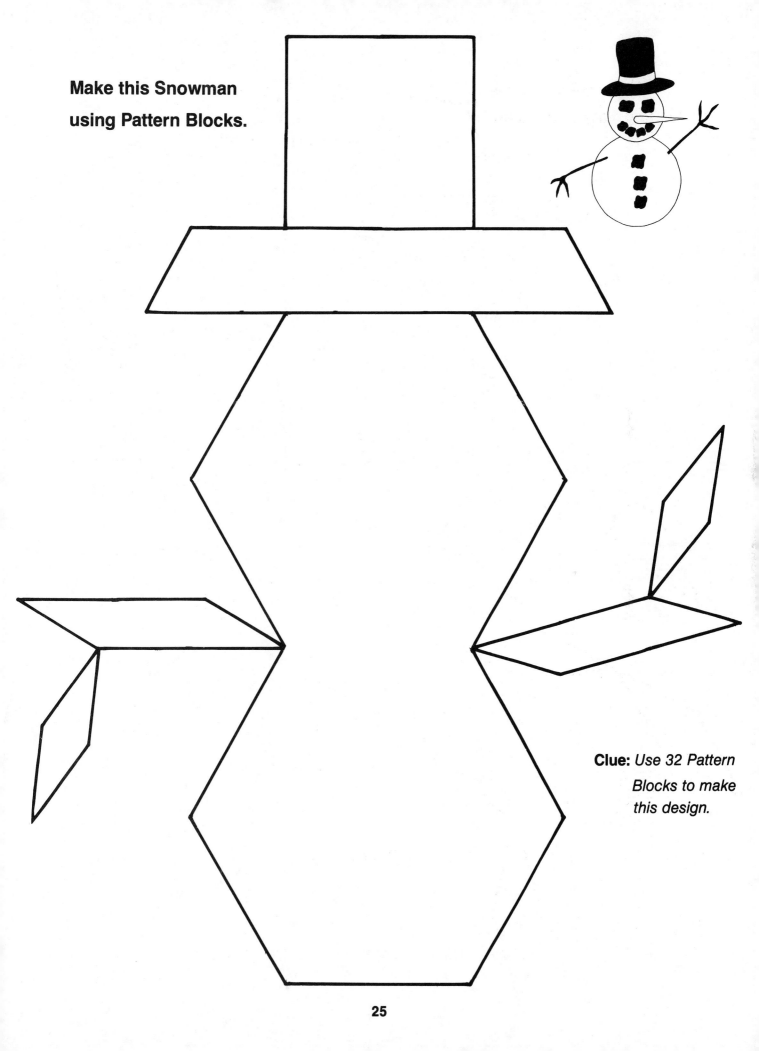

Clue: *Use 32 Pattern Blocks to make this design.*

Can you make this Camel using Pattern Blocks?

Clue: *Use 5 blue pieces and 7 red pieces.*

Use Pattern Blocks to make this Mother Bird.

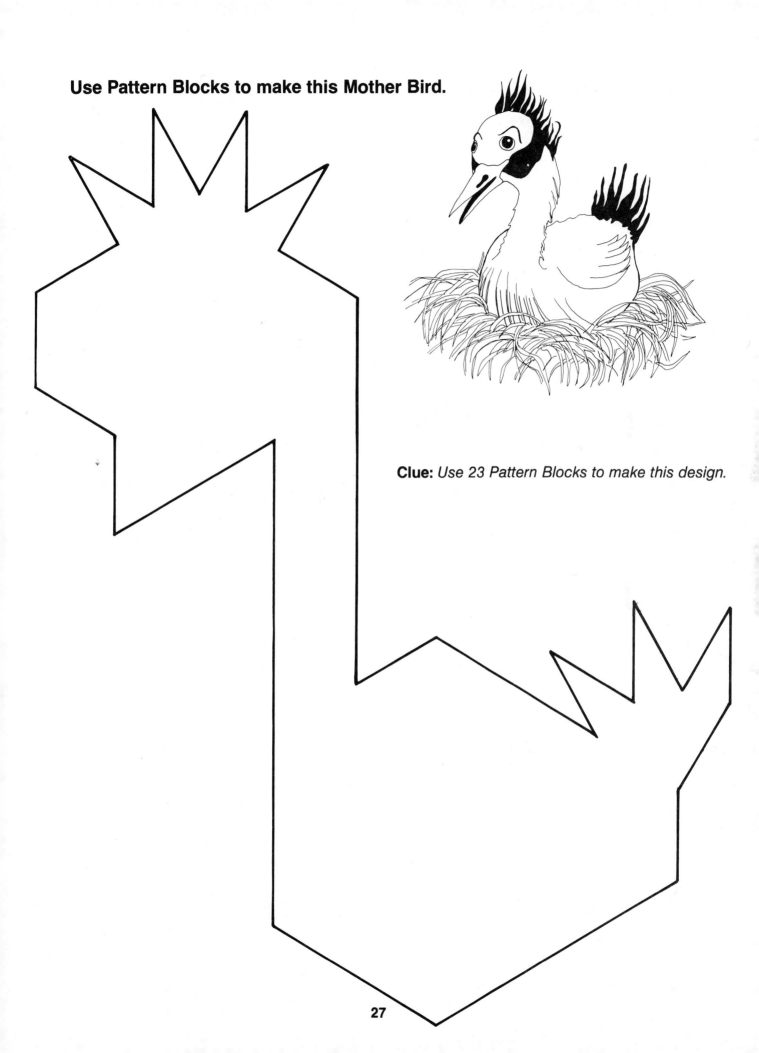

Clue: *Use 23 Pattern Blocks to make this design.*

27

Use Pattern Blocks to make this Dinosaur.

Clue: *Use 6 yellow hexagons.*

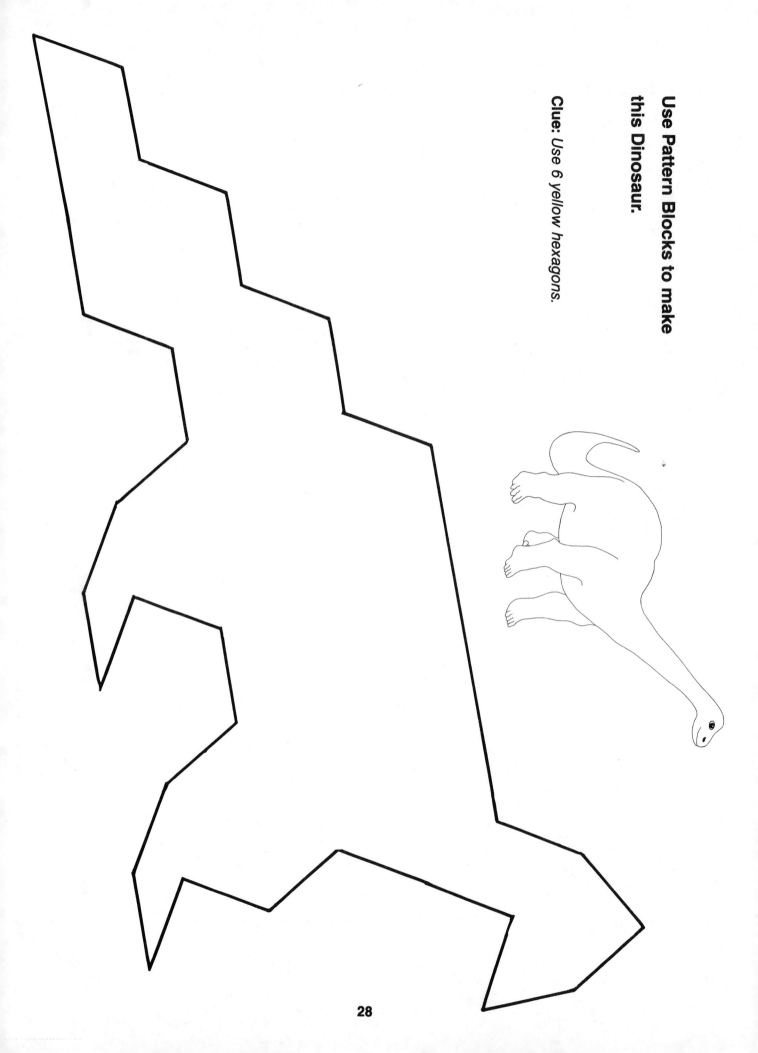

Make this Boat using Pattern Blocks.

Clue: *Use 6 green pieces and 7 orange pieces.*

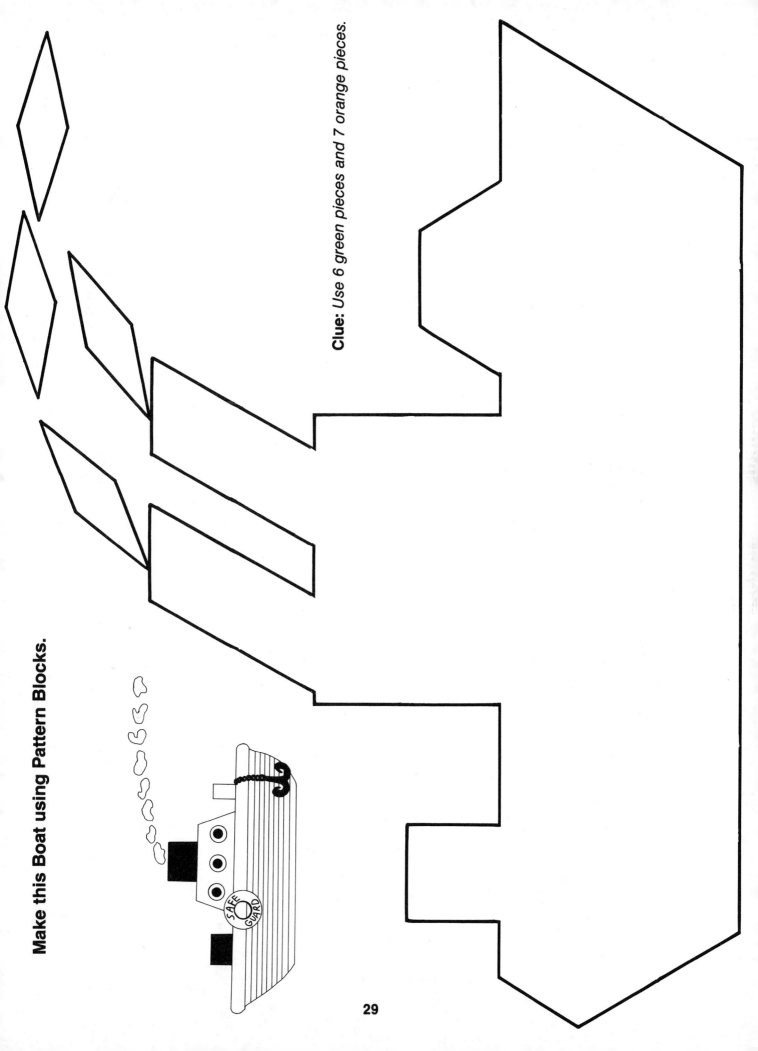

29

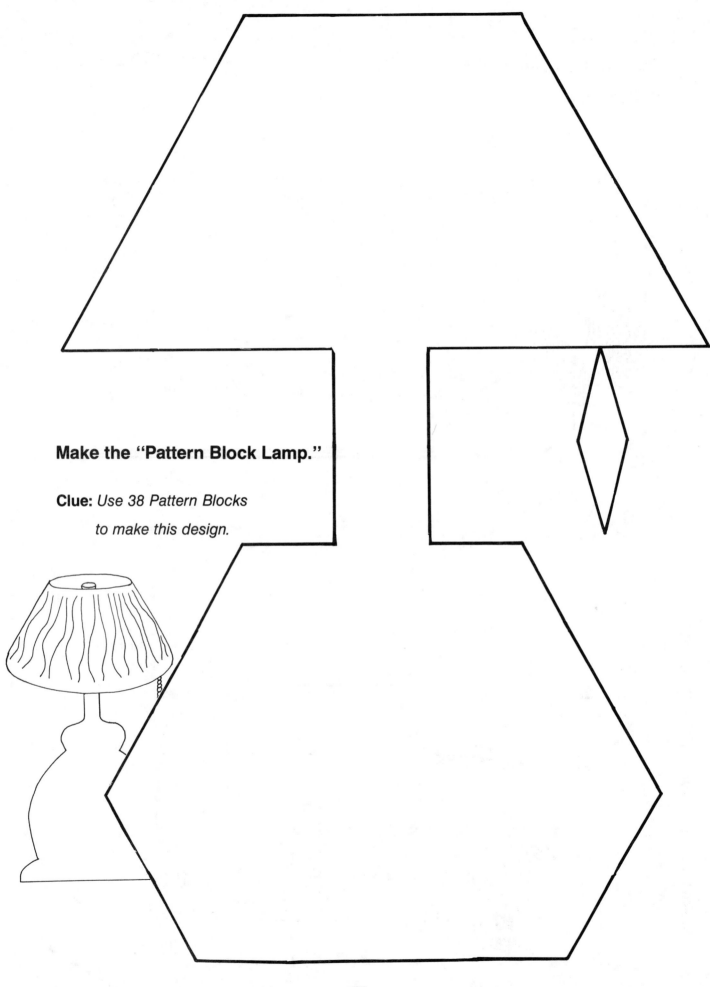

Make the "Pattern Block Lamp."

Clue: *Use 38 Pattern Blocks to make this design.*

30

Build this Rocket using Pattern Blocks.

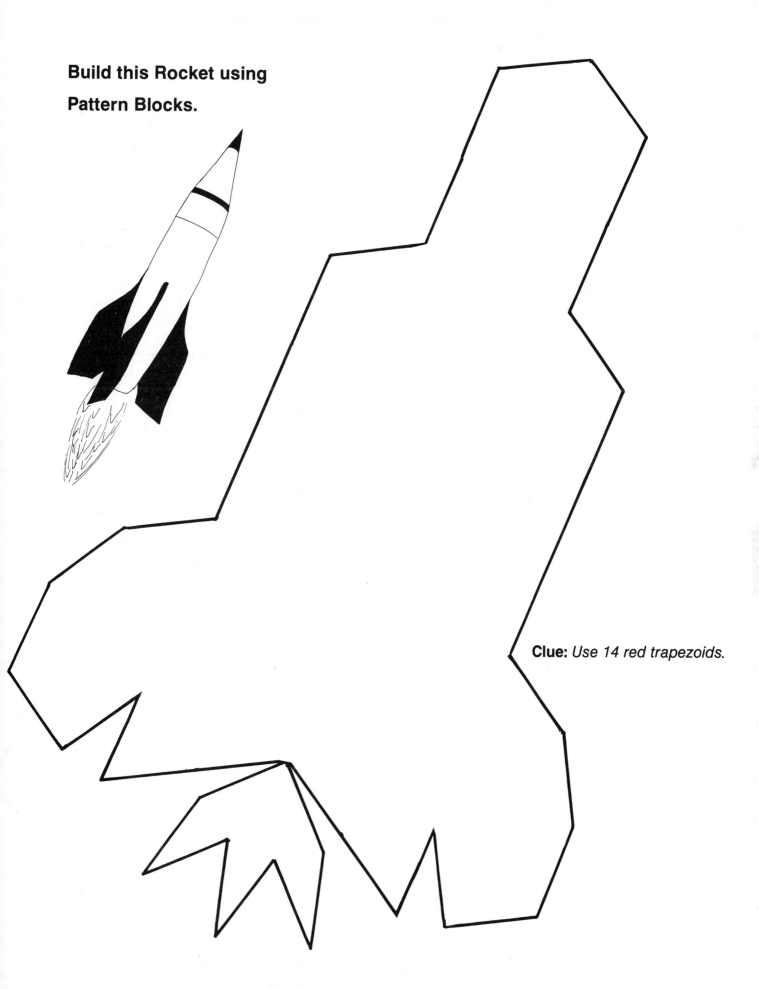

Clue: *Use 14 red trapezoids.*

31

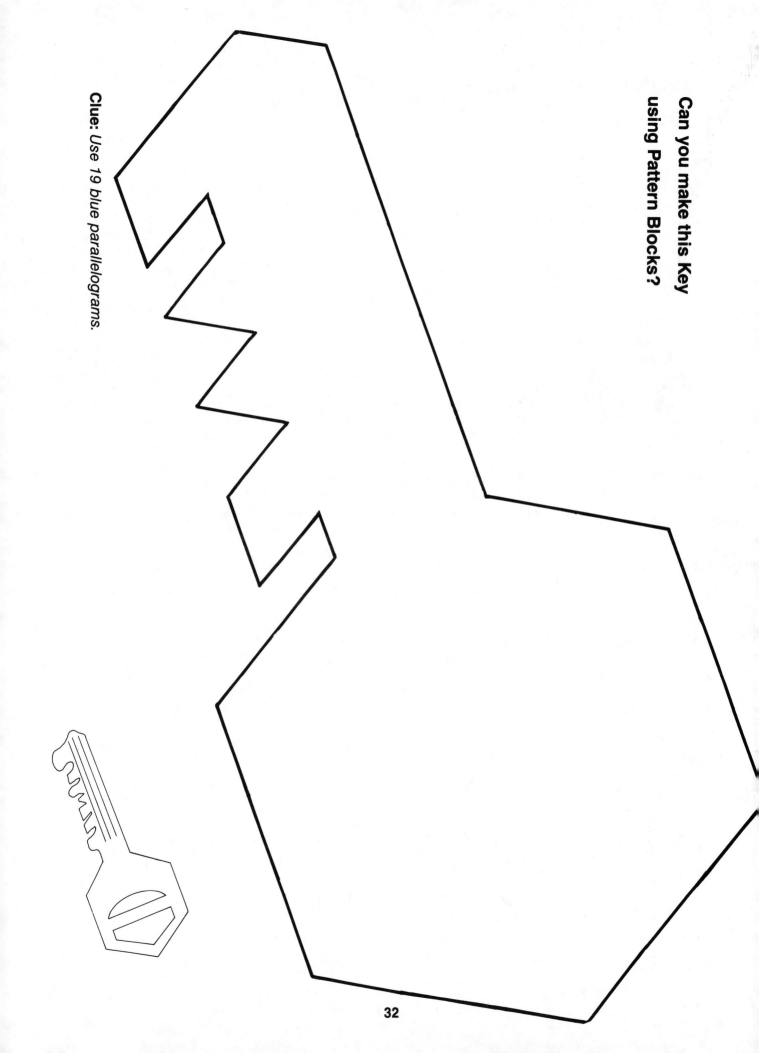

Can you make this Key using Pattern Blocks?

Clue: Use 19 blue parallelograms.

**Use Pattern Blocks to make
this Holiday Tree.**

**You will need Pattern Blocks to
make this Kangaroo.**

Can you make this Cat using Pattern Blocks?

Make this Lamb using Pattern Blocks.

Make this "Pattern Block Train".

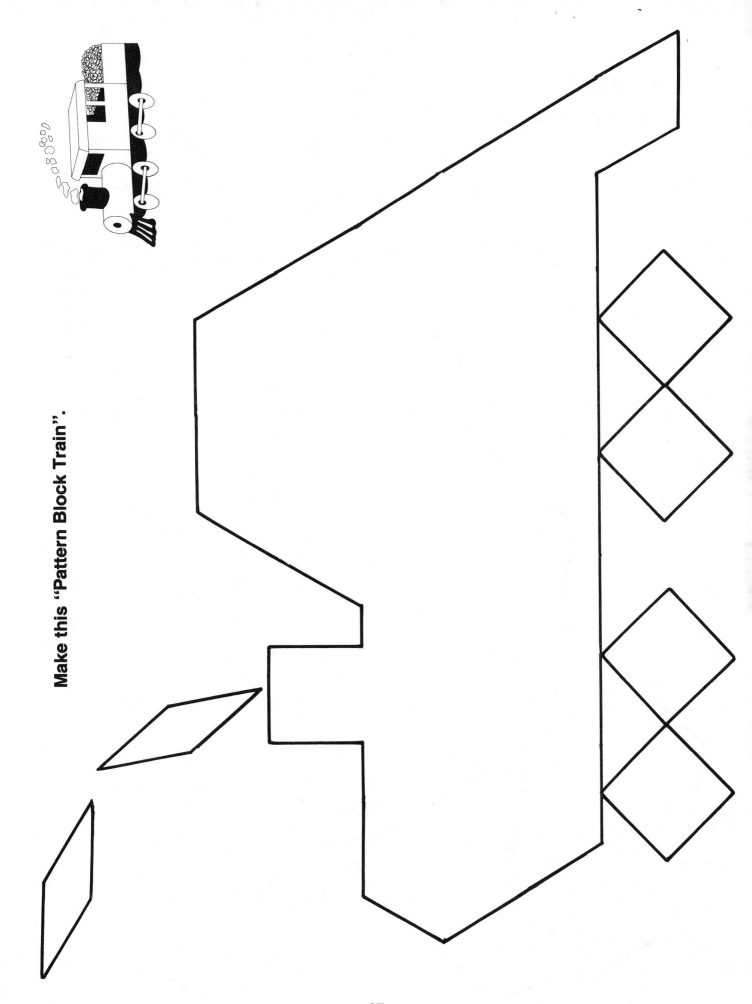

37

Make a Guitar using Pattern Blocks.

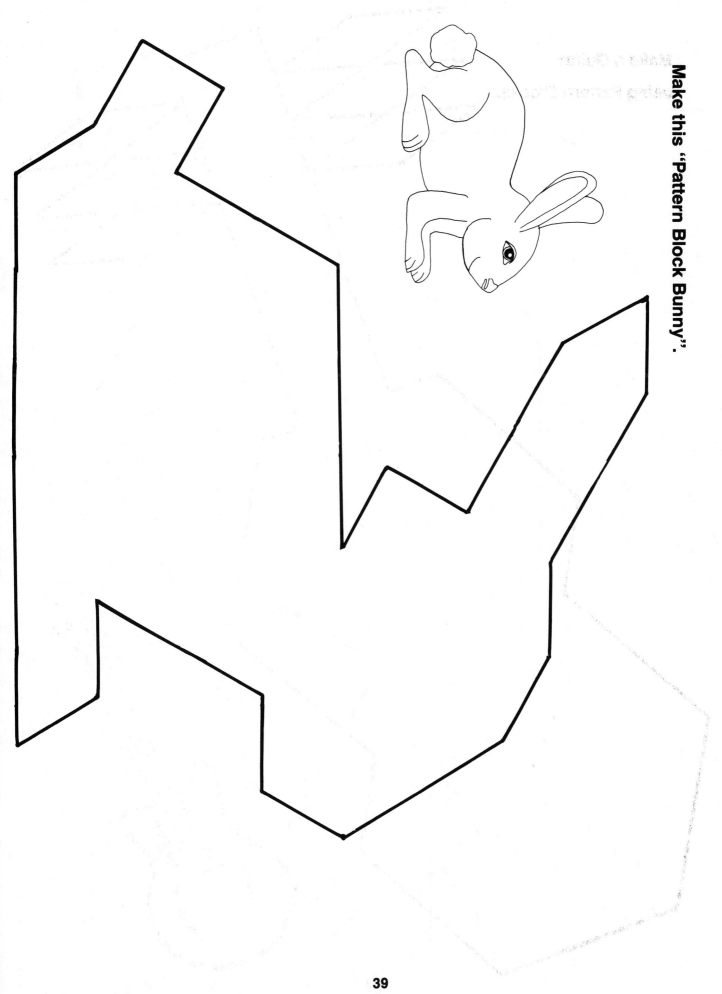

Make this "Pattern Block Bunny".

Can you make this Butterfly using Pattern Blocks?

41

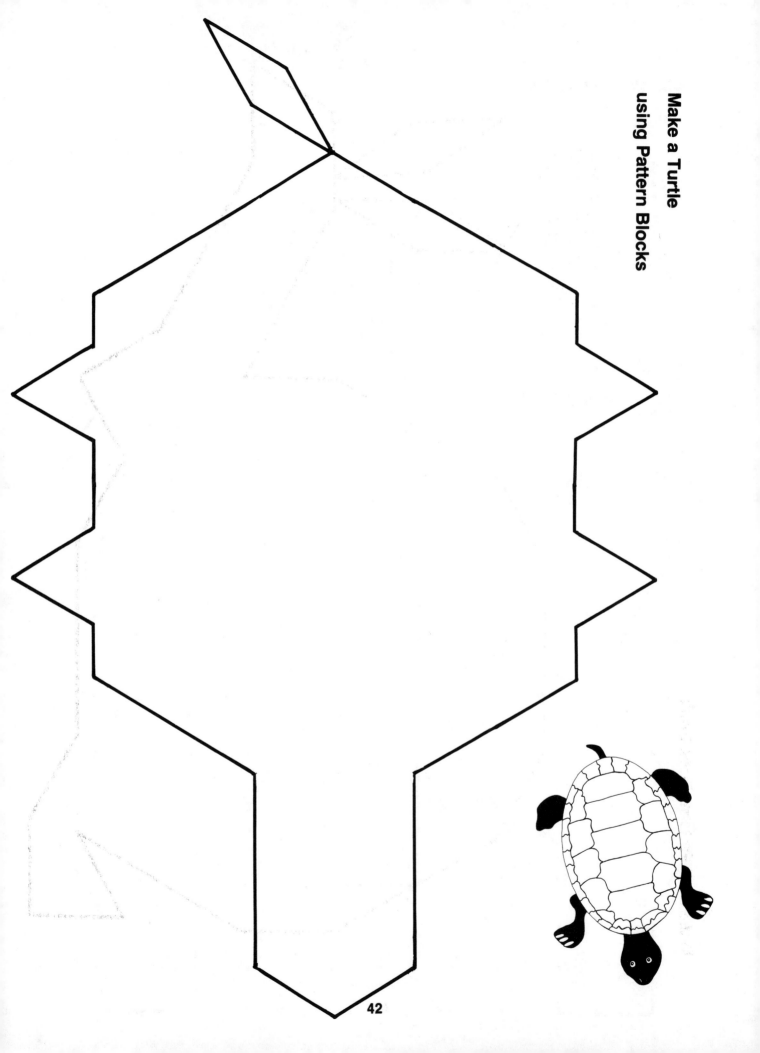

42

Make the "Pattern Block Snail"

Make this Jumping Frog using Pattern Blocks.

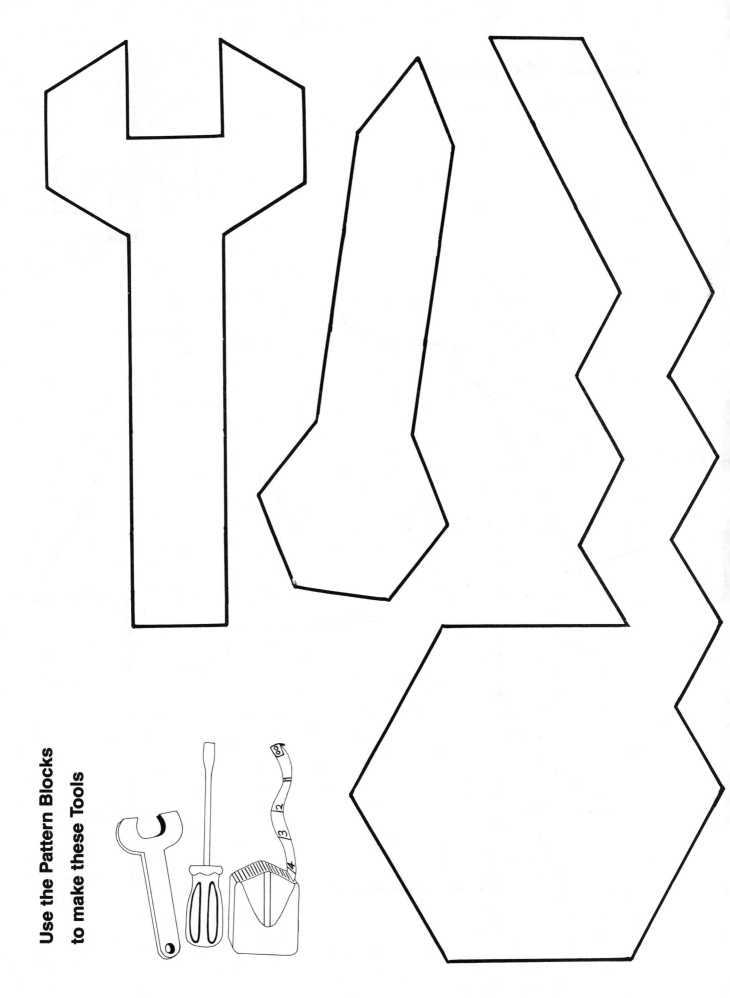

Use the Pattern Blocks to make these Tools

45

Make this "Pattern Block House."

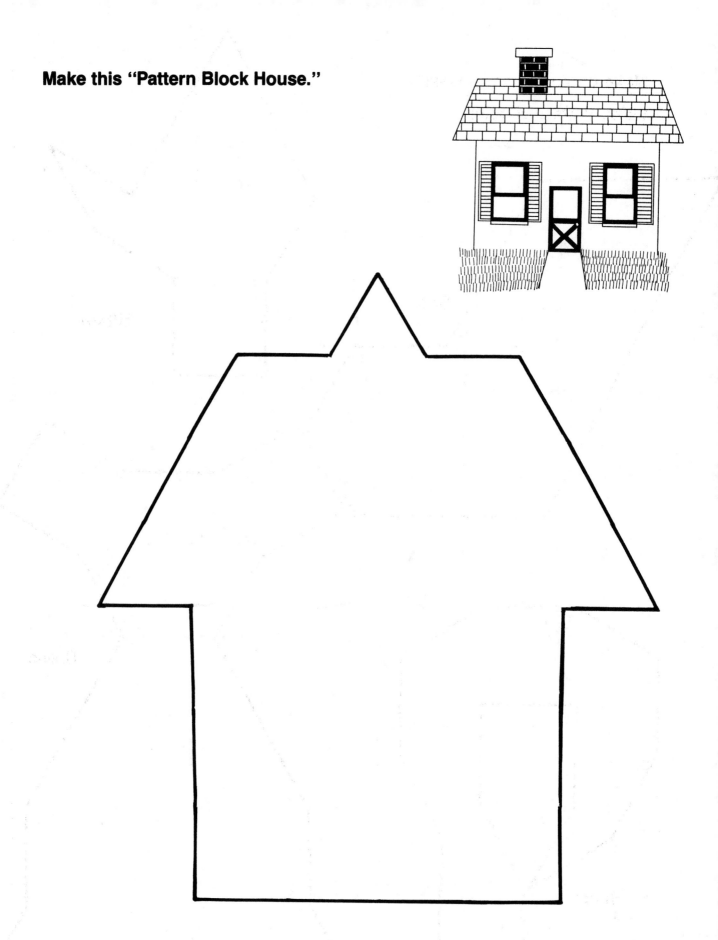

Make these things to wear using Pattern Blocks.

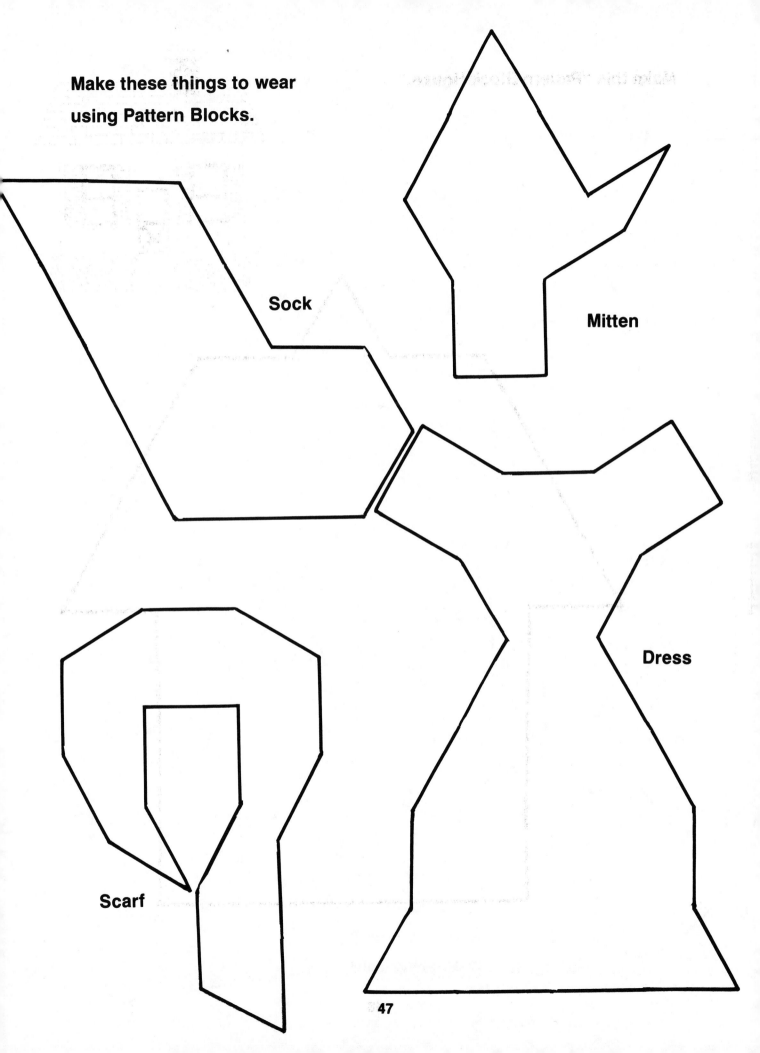

Sock

Mitten

Dress

Scarf

47

Make these shapes with Pattern Blocks.

Trace the shapes on paper.

Color each shape so that it looks like an amazing animal.

Make this shape using 5 Pattern Blocks.

Make it again using 10 Pattern Blocks.

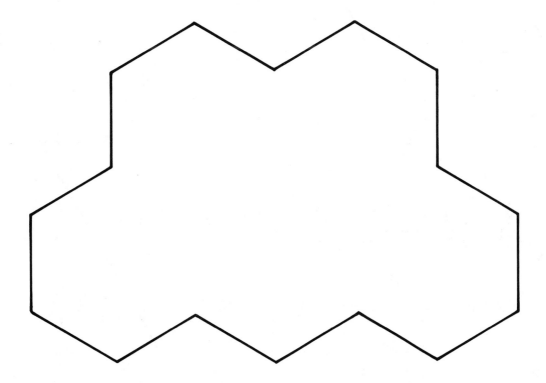

Make this shape using only red and blue Pattern Blocks.

Then make it using only yellow and green pieces.

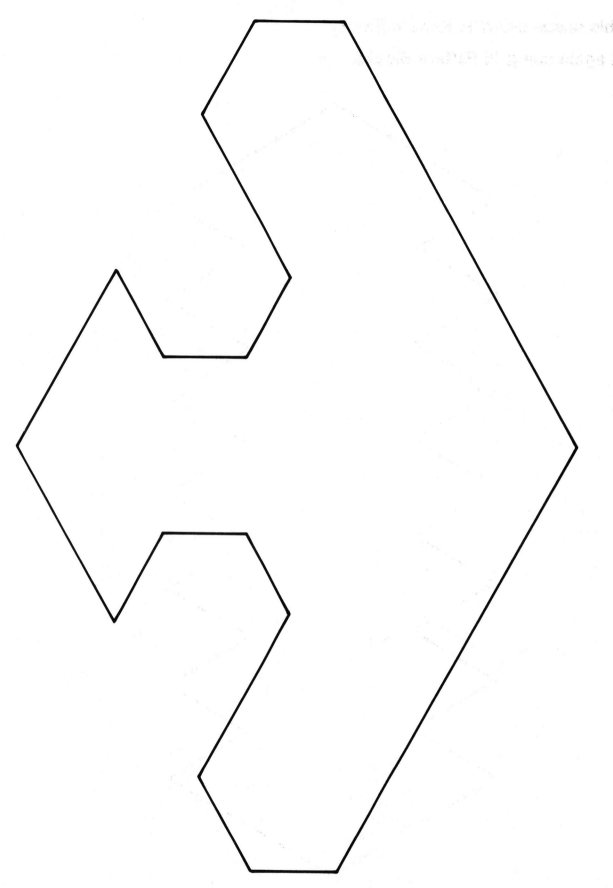

Make this shape using 18 Pattern Blocks.

Make it again using 15 Pattern Blocks.

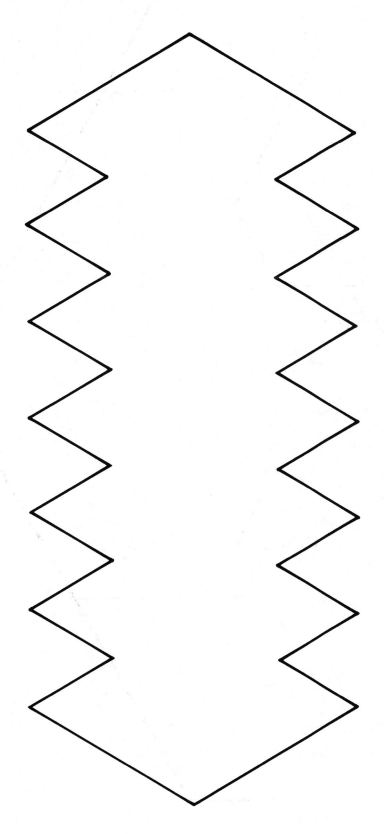

Make this shape using the greatest number of Pattern Blocks.

Then make it using the least number of pieces.

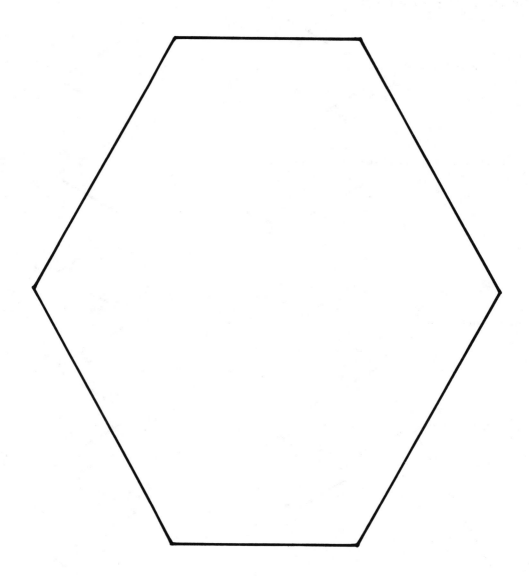

Clue: *The greatest number of pieces equals 42.*

The least number of pieces equals 42 –32.

Complete the pattern so that the dotted line becomes a Line of Symmetry.
The bottom part of the design should be a mirror image of the top part.

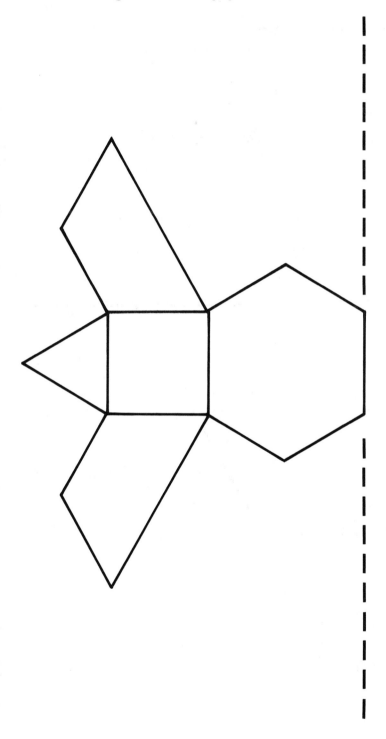

Can you find other Lines of Symmetry in your designs?

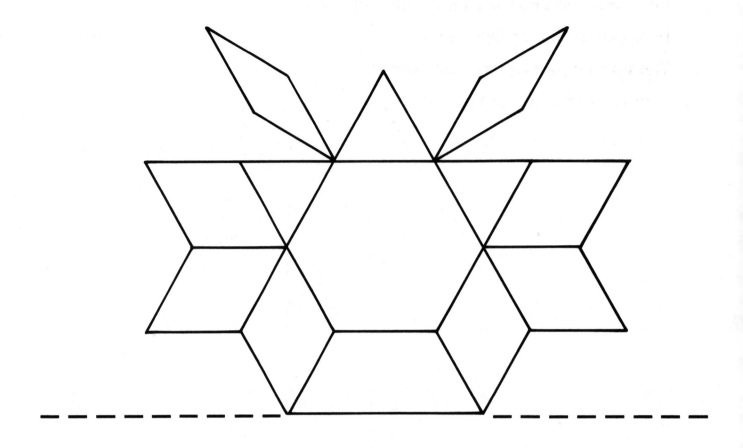

**Complete the pattern so that the dotted line
becomes a Line of Symmetry.
The bottom part of the design should be
a mirror image of the top part.**

Complete the pattern so that the dotted line becomes a Line of Symmetry.

The bottom part of the design should be a mirror image of the top part.

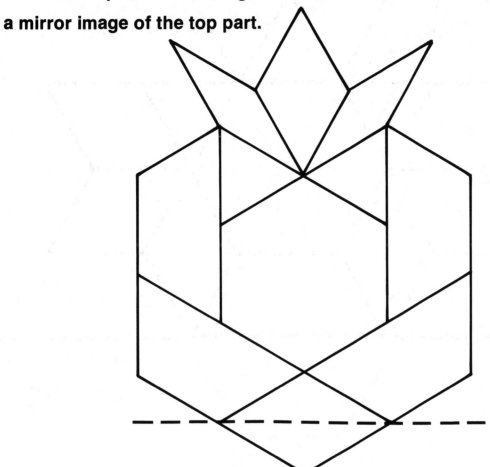

Can you find other Lines of Symmetry in your designs?

**Complete the pattern so that the dotted line
becomes a Line of Symmetry.
The bottom part of the design should be
a mirror image of the top part.**

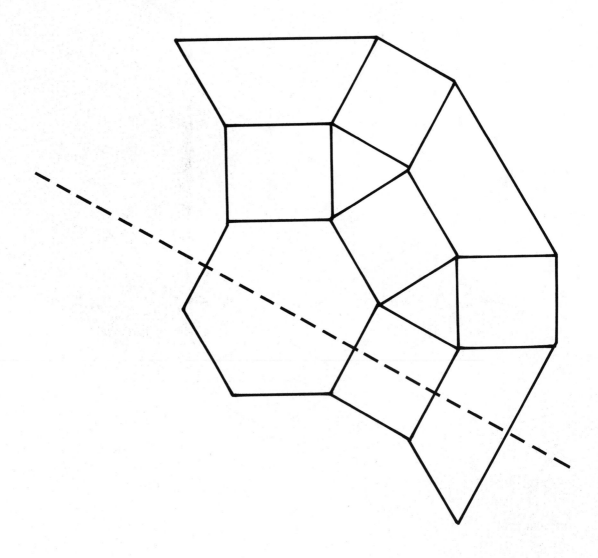

**Can you find other Lines of Symmetry
in your designs?**